KANBAN

The Beginner's Super Project Management
Guide

(How to Visualize Work and Maximize
Efficiency and Output)

Ronnie Hames

Published By Ronnie Hames

Ronnie Hames

All Rights Reserved

Kanban: The Beginner's Super Project Management Guide
(How to Visualize Work and Maximize Efficiency and
Output)

ISBN 978-1-77485-416-7

Legal & Disclaimer

The information contained in this book is not designed to replace or take the place of any form of medicine or professional medical advice. The information in this book has been provided for educational and entertainment purposes only.

The information contained in this book has been compiled from sources deemed reliable, and it is accurate to the best of the Author's knowledge; however, the Author cannot guarantee its accuracy and validity and cannot be held liable for any errors or omissions. Changes are periodically made to this book. You must consult your doctor or get professional medical advice before using any of the suggested remedies, techniques, or information in this book.

Upon using the information contained in this book, you agree to hold harmless the Author from and against any damages, costs, and expenses, including any legal fees potentially resulting from the application of any of the information provided by this guide. This disclaimer applies to any damages or injury caused by the use and application, whether directly or indirectly, of any advice or information presented, whether for breach of contract, tort, negligence, personal injury, criminal intent, or under any other cause of action.

You agree to accept all risks of using the information presented inside this book. You need to consult a professional medical practitioner in order to ensure you are both able and healthy enough to participate in this program.

TABLE OF CONTENTS

Introduction

Kanban is a method that is employed by the majority of organizations to help visualize and manage bottlenecks or work-in-progress in any project within the organization. Kanban loosely refers to a the signboard or signal card in Japanese. This device was created in the 1940s , by Taiichi Ohno to enhance the efficiency of Toyota.

Toyota sent a note card to the supplier in order to inform them that they required more of a particular part. the same card was glued onto the component after it was shipped to Toyota. After the components were used the card was given to suppliers and manufacturers to inform them that the item was required again. This made it easier for the company to put together the parts in time and then send them to customers.

The same procedure is applicable to various companies in diverse sectors.

Kanban boards, as well as Kanban software can help a supervisor or the entire team to see the pace at which a project is moving. With these tools, they are able to recognize obstacles and find methods to get rid of them. Through the book, you'll learn more about the basics of what Kanban is and the ways it can be applied to enhance the productivity.

There are various principles and methods that an company must follow when setting up Kanban systems. It is essential that the team's manager and team members are aware of these principles and operate effectively. There are many benefits to making use of the Kanban tool in the workplace as it boosts productivity and improves the quality of work.

This book will assist you to learn the ways Kanban methods can be implemented in various departments of the organization and how they could be applied in various sectors.

Kanban can also be used to forecast the results of specific processes by using statistical tools , such as charts and models. This is discussed in the final chapter in the book.

The benefit of Kanban is the fact that it allows to increase visibility within the workplace since each process is seen as a picture and every task in that process is tracked and completed.

Kanban is also a method to enhance individuals' work ethic.

Chapter 1: A Brief Introduction Kanban

Kanban is a method created to help manage the work that is moving through the process or progress. With the help of boards in the management system it is possible to visualize the workflow process as well as the actual work going through the process. The aim for the software is to find possible bottlenecks or obstructions to the process and correct them to speed up efficiency.

A Short Histories

Taiichi Ohno Taiichi Ohno Japanese Industrial Engineer and Businessman for Toyota automotive, developed Kanban in the 1940s. The system was designed to assist teams in planning their activities and deal with any problems that arise in the course of work. The software was designed to facilitate the work process at each stage that the procedure.

Kanban was invented because Kanban was created because the Toyota company in Japan was not performing in the same way as the competitors in America. Utilizing Kanban, Toyota achieved control systems that increased their efficiency. Kanban also assisted in reducing the amount of wasted raw material utilized.

A Kanban system regulates the value chain from the supplier of raw materials to the end user. This can prevent interruption in supply and decreases any excess of goods in different phases during the course of.

This software demands constant monitoringand attention should be paid to spot bottlenecks in the process and prevent these from the beginning to speed up the process. Every company strives to reach an efficient output while reducing time to deliver. In the past, Kanban has been used in various systems to boost efficiency.

How do I use the Kanban Method?

Kanban is a method of organizing and managing information. Kanban method was developed in factories by Taiichi Ohno. It was later adapted to the IT industry knowledge work, software development and knowledge development in 2004 , by David Anderson. David utilized the work of Eli Goldratt, Peter Drucker, Edward Demmings, Taiichi Ohno, and many others to establish the process using concepts such as queueing theory as well as pull systems and flow.

Kanban Change Management Principles

The Kanban method is based on a set of guidelines and practices that aid in improving and control the flow of work. It is non-disruptive and advancing. It assists in improving the processes that are being worked on within an organization.

Your company will achieve the Kanban principles and practices Kanban for maximum value of any business process, through improving flow of work, reducing the time for delivery and increasing satisfaction with customers.

The basic principles and methods of the method are described below.

Fundamental Principles

Always begin with what you are doing right now

The Kanban method doesn't necessitate making any changes to the process immediately. Instead, you can rely for your workflow, and adjust the flow as required and over time. It is important to make sure that your team is in agreement when you make changes to the procedure.

Explore incremental and evolutionary change

Kanban is a system that Kanban system encourages each team along with its team members to adopt minor adjustments instead of making major changes that could cause opposition from those in the team as well as the company.

Be aware of current roles, responsibilities and Designations

In contrast to other methods, Kanban does not impose any modifications to the structure of the company. It is not necessary to change any of the current roles and functions that aren't performing adequately. With Kanban the team will determine the necessary changes and then implement the changes as needed. The three basic principles can aid an organization in overcoming the fear of changing.

Encourage Actions of Leadership

Because the Kanban system is designed to encourage constant improvement at all levels of the company the leadership actions don't require only to be done by top management. All employees are empowered to lead and find ways to improve the procedure. It is encouraged for them to think of ways to improve the way the products and services are provided.

Core Practices

Visualize the process

If you decide to adopt and utilize this Kanban method, you need to visualize the process. You should visualize the steps on the physical board or on a Kanban board to visualize the steps required to complete work or other service. A Kanban board could be straightforward or intricate based on the various kinds of work items teams must be working on and deliver.

After the process has been visualized, it is now time to visualize the work currently being performed on behalf of the entire team. This can be done with cards of various colors or notes to highlight the various kinds of work that are being done on behalf of the entire team. It is possible to include various columns and colors to identify the procedures and work which is performed by each person on the team and the whole team. These boards are able to be modified to enhance the efficiency of a procedure.

Limit or reduce the amount of work-in-progress

It is crucial to restrict work-in-progress in order to use Kanban. Kanban system. This will motivate your team to finish the tasks that are ongoing prior to starting any new task. This means that your team will only be able to begin the new project when the work that is currently in progress is marked as complete. This will increase the capacity of the team to take on more work.

It could be difficult to recognize the limit of your work-in-progress. You can begin with no limits on your WIP or Don Reinertsen had suggested that teams should begin with no limits on their WIP and then watch the performance of the team once you have begun using Kanban. When you have enough information to establish the WIP limits at each step in the process. The majority of teams begin with a WIP limit which is 1 or 1.5 times the number of employees in the group working on a particular task.

It's beneficial for the team to restrict WIP and set WIP limits as the team members will complete their work before moving to new projects. This will also inform clients and stakeholders they have a restricted capability of work that could be completed in the group. Therefore, careful planning should be considered when a new project or request is presented for the group.

Control the flow

It is vital to manage and improve workflow after you've completed and are proficient in the first two tasks. A Kanban system assists the team to control workflow by showing different phases of the workflow as well as the status of work that must be completed at every stage. Work in progress limits for WIPs are established according to the way the workflow is defined. Anyone in the team can see that work is completed in time, within the WIP limit or the work is being piling up due to one project that is being getting delayed. This impacts how fast the team is able to complete work.

Kanban assists teams in analyzing the system, and makes changes to the system in order to enhance efficiency. Kanban can also help teams see how they can decrease the time it takes to finish a task. It is crucial to observe the stages of wait between tasks to see how work is being done, and to identify and eliminate any bottlenecks in the process.

A team must examine how long they're in the stages of intermediate wait and determine ways to decrease the amount of time they spend in these sections. This is a vital aspect to take into consideration when a team is trying at ways to speed up the cycle.

When workflow is improved the team's performance becomes more predictable and seamless. If it's reliable, it's easier for the team to fulfill the commitments they made to clients and finish the task within the deadline. It is crucial for the team to find the method of forecasting the time to complete work, since it can improve the efficiency of the team.

Create clear policies

To visualize an entire process, it is beneficial that a group establish and visualize policies that describe how work is completed. The team manager or project manager should create a policy that defines the manner in which work is to be done within the system. Policies can be created at a board level for each column, or at the swim lane level. The policies may contain checklists that should be checked off at each stage of the process. The checklists should be prepared for each task completed by the team members to assist team members in managing the flow of work. For example, the description of when a job is considered complete or the explanation of when tasks may be pulled or pushed should be put in guidelines.

Feedback Loops

Feedback loops are an essential component of any audio system, and the Kanban system encourages all teams to create various positively feedback loops. Every team has to examine different phases of the workflow using Kanban boards, reports and metrics. Kanban board, and report as well as metrics which can be utilized to improve any process.

The phrase "Fail frequently and quickly," may not always work for all teams and is not often understood by the majority of teams. But, if a group gets feedback at a very early stage of the process and the team or director will have the ability determine a method to complete the project on time with fewer mistakes. A feedback loop is essential to ensure.

Develop and improve collaboratively and in a controlled manner

Kanban is a method of improvement that allows teams to make small changes before making big improvements Kanban technique is an enhancement technique where teams can make small adjustments to the procedure prior to making major changes. This allows teams to manage the changes swiftly. This method allows for using statistical techniques that allow teams to construct an hypothesis and then tests the hypothesis to discover the results.

The most important task for teams is to assess and modify a procedure whenever needed. The effects of the modifications made can be assessed and observed by observing various indicators on the Kanban board. These signals assist the team assess whether the modification made to the process is actually improving the process, and determine whether they want to keep the change.

A Kanban system can help the team gather important data about the performance of each person on the team as well as the entire team. This information will assist the team develop metrics which can be used to measure the effectiveness and adjust the system as required.

Is Kanban dead?

There are many who are unaware of Kanban and even more those who would tell you that Kanban has become obsolete. But, this is not the case. Kanban is pull-based system, which, in contrast to push system, can be a good system. Kanban systems allow you to monitor and manage the work of different departments of an organisation.

New methods are being created to make it easier for managers to monitor the work being performed by teams within an company. It is true that businesses that are lean are shifting towards a shift away from Kanban and adopting newer techniques and tools to keep workflow. It's also true that the majority of methods that are being developed employ the principles and methods of Kanban.

So, is Kanban dead? It isn't. There are many developments happening in the world of lean delivery which increase the efficiency of an organisation in general. Kanban is one of them. Kanban is a tool that is changing the face in lean delivery.

Chapter 2: What Is Kanban Work?

Kanban is an approach to managing change that is both adaptive and non-disruptive. That means that processes already in place are improved through small changes. The risk to the entire system is lessened by making small changes rather than large ones. This strategy results in little or no resistance from the stakeholders, team members and the customers.

In the previous paragraph the team needs to visualize the process and process flow. This can be accomplished with the Kanban board which is comprised of a whiteboard with some sticky notes with different hues. Every note that is on the board is utilized to represent a particular job.

Three columns are featured in the traditional Kanban board.

* "To Do" Tasks which aren't yet completed and may be delayed

* "Doing" What are the tasks which are being completed by the team

* "Done" Tasks that were completed

This type of visualization can lead to clarity since the allocation of work is clear and the bottlenecks in every task are identified at the beginning. Kanban boards are able to show intricate workflows, based on how complicated the workflow is as well as which components of the workflow have visualization. This will help the team to eliminate any bottlenecks in complicated processes as well.

Concept of flow

The basis principle of Kanban's Kanban process is the idea of flow. This implies that the cards that represent processes should flow through the workflow as smoothly and evenly as is possible without any blockages or waiting periods. Any potential risk that could slow working efficiency must be analyzed carefully and the Kanban system is a variety of techniques such as models, models and metrics that can be employed. This system assists teams to attain kaizen, which is an ongoing improvement.

The concept of flow is essential to a process. If a team is able to see the flow and works towards improving it, speed of delivery could be increased rapidly. This can aid the team in reducing duration and improve the quality of work by receiving rapid and constructive feedback from customers, stakeholders, or other stakeholders.

Kanban WIP Limits

A majority of teams and their members tend to multi-task as they are pushed to complete the most tasks they can within the time frame. The Kanban system can help reduce multi-tasking and emphasizes the principle "Stop making a start and get to work."

As we mentioned previously the team should determine WIP limits at each stage of the Kanban board, and also encourage participants to complete their current tasks before moving on to new tasks.

Chapter 3: Kanban Systems: Types Of Kanban Systems

Numerous companies have begun using Kanban methods to boost their efficiency. There are a variety of cards that are able to be utilized in Kanban systems. Kanban system, since Kanban unlike 6-Sigma is not a set methodology. This is why it is that it can be utilized for a variety of uses. The various types that comprise Kanban systems are described in this chapter. This isn't an exhaustive listing of all the Kanban system. It could mean various terms to various companies.

Production Kanban

The system comprises an extensive list of tasks which must be carried out in order to ensure that the product is delivered within the timeframe. The system collects details about the different materials and components that are needed together with information from the withdrawal systems. This system allows the team to begin in the process of producing the product, and then provide the information on the services or products required to be created.

Withdrawal Kanban

This system is also referred to as a transfer or move cards Kanban. When a component requires to be moved to the manufacturing Kanban to another kind it is used for signalling. The cards are tied to various tasks which must be transported to the workplace at the time they are due to be completed. After the task has been completed and the cards have been returned, they are returned.

Emergency Kanban

This type of system can be employed to replace malfunctioning parts or to announce to all the team members that the amount of a service or product needed to be produced has increased or decreased in size. Companies often employ an emergency Kanban systems when a certain component of the system is not functioning as it is supposed to or if there are any changes to the procedure.

Through Kanban

Through Kanban Systems include both the production as well as withdrawal Kanban systems. These systems are employed when both workstations that are part of both Kanban systems are close to each other.

This system accelerates the production process. For instance when an organization has its areas of production as well as the area of storage adjacent to each other The system will draw components from both systems and move these parts as they move through that production line.

Express Kanban

This type of system is activated in the event of a deficiency of components within the system. The systems transmit signals to teams to boost the quantity of components required to finish the task at the first place. This system is intended to make sure that the manufacturing or process of production is not slowing down. They are also known as signals Kanban systems as they can be utilized to trigger every shortage, or to trigger purchases.

Supplier Kanban

Suppliers are an entity or person from whom an company sources materials to create its products. The system is directed directly to the supplier and is typically used as a representative for the company that manufactures it.

Whatever the kind of Kanban system is used regardless of the type, it is crucial to remember that the Kanban system can be used to improve the efficiency and the quality of products and services offered by an business.

Chapter 4: The Benefits Of Kanban

The idea of managing the creation of the product is not something that is brand new. The underlying philosophy behind Kanban is widely known since its inception during the 1940s.

The principles behind management of product development and creation of knowledge on the two Taiichi Ohno's Toyota Production System (TPS) and Lean Manufacturing originate from Kanban.

If a company decides to control work instead of managing its employees, this will help to create a healthy humane, creative and a synergistic environment for people to realize their full potential. This is an easy method to bring about changes in the company. This method isn't intrusive and has been accepted by a variety of new companies because it starts by directing the tasks that are performed within the company.

Let's look at ways Kanban can benefit your company.

Business Values are the first priority

David Anderson who is considered one of the original pioneers of the Kanban method has said that Kanban isn't just an approach or system to control the progress of projects but it is also a framework for decision-making which is extremely efficient. The system was designed in a manner that permits organizations to make decisions specific to certain objectives in the field of economics.

Because organizations operate in a competitive environment they have to prioritize, execute and complete tasks as fast in the time they are able to. They should also be striving to ensure that their work is error-free , to stay an inch ahead of their rivals.

Don Reinertsen has said that companies should always calculate the expense of putting off the release of a product. In addition, the business might also consider determining the costs of applying one aspect over another might be, as the feature being introduced could represent the feature that makes the business apart from other in the same sector. It is also necessary to calculate the time it takes to use one particular feature over the other.

Enhances Visibility

Much of the work carried out in the organization takes place hidden from view and it is vital to present that work to the top executives of the company and the customers of the organization. This is the primary function of Kanban. Kanban boards can serve as a source of information for observe the progression of a procedure and any bottlenecks or obstacles during the process from a glance.

The information is accessible not just to members of the company, but as well to all interested parties, customers, or observers. This allows for the sharing of boundaryless information throughout the company. The limitations of the software aid in understanding the way in which work is assessed and the way it is being ranked.

Reduces Context-Switching

The amount of work tasks taken care of by employees is counted within the Kanban system. This number is reduced by the WIP limit. This system is designed to help employees to complete tasks that are both important and valuable, which enhances its value to your company.

Kanban helps teams avoid overloading their members with personal WIP limit. Each member is only able to start another project when the member has completed the task they've pledged to complete. The Kanban list is based on the principle : 'Stop beginning and begin working on finishing. This system has helped teams to focus and deliver their downstream activities successfully.

Improves Collaboration

In the majority of organisations, departments are separated from each other. There are instances where there are conflicts that occur between the teams that deliver software and the product management. With Kanban the teams are part of the value system for development.

Kanban is an e-pull system that promotes synergies. It also helps break down the barriers that separate departments and specializations. This can create cross-departmental collaboration. The process of transferring tasks to Kanban boards gives teams the chance to share their knowledge and collaborate with each other.

Reduces the wasteful activities

A lot of project managers concentrate on timelines rather than process queues as the latter is deeply embedded into their minds. They utilize Gantt charts and other forms to aid in assessing the timelines for each person on the team. The most common mistake project managers make is to grasp is that they have to accept the possibility of uncertainty, not only specific planning. Many project managers attempt to create activities that add to the duration of the project and the risk involved and thus increase the number of process queues.

The Kanban board is a way to strengthen some WIP limitations that makes it an apuls-based system that lets a team or organization to have a consistent supply of top-quality ideas which can be completed in the appropriate timeframe. This helps eliminate any work considered unnecessary, thereby reducing the number of queues to be built.

Some of the activities upstream like workshops, gathering requirements and business cases are conducted in the event that they are required. The project director has to take decisions in a timely manner. Kanban permits an administrator to oversee multiple projects that are relevant to validating and prioritizing concepts across the board.

Introduces True Sustainability

Kanban methods help teams organize their work at a steady, sustainable, and seamless pace. This helps reduce stress, frustration and inability to commit, thereby raising the likelihood of employee turnover. Limits on WIP help control the speed of a process, which encourages creativity.

The teams will not commit to a procedure in the beginning and abandon that commitment in the future. Teams will be able to explore and solve problems using different methods to come up with solutions that are less prone to quality concerns.

Enhances Quality

For most professionals, the initial quality is an essential aspect of a successful project. The impact of errors can be devastating to the productivity of the team. They also assist the team to address any quality issues from the beginning that can improve productivity. There are many tasks, such as collaborative analysis, user documentation, that help to create top-quality software.

The policies of Kanban software help to establish professional standards , which are endorsed by various boards, including project managers products, software items, managers as well as business stakeholders and customers. The software clearly specifies these guidelines in every step of the process.

Increases Morale

The Kanban software allows many agile and traditional teams to shift away from pressure to commands and control methods. Each member of the team is required to create their schedules according to the workflow managed by the system and not their supervisors. Each member of the team is expected to complete their work at a quick but steady and sustainable speed, which can help induce a positive stress, known as stress. This allows the team to see the fruits of their work. Because Kanban oversees the activities of each member of a team, members will not feel anxious, and they can concentrate on their tasks and unleash their talent and creative abilities.

Instills Kaizen Culture

Kaizen board Kaizen board gives team members and managers with queues that can be controlled and need fewer buffers. Pull-based systems help identify any problems with the process, bottlenecks obstacles, delays, issues with delivery or miscommunication, agency problems, inefficiencies and confusion.

These queues bring attention to problems and assist members and teams identify solutions prior to development slowing down. By using groups, Kanban introduces Kaizen, which is a philosophy that lets the team is taught to constantly improve the way it deals problems all over the board.

Introduces Predictability

Kanban was believed to be a radiator of information during the time of its development. By using lists and queues that are displayed on the board, Kanban can help a manager anticipate the majority of issues based on current or historical information. This minimizes the risk of making a guess on the part of the project manager. If improvements are made from actual data, time to deliver and lead times are improved with time.

There are numerous benefits of making use of the Kanban system. A company will be able save money and improve efficiency by using this Kanban system.

But, there are a few points to keep in mind prior to installing the system.

The team needs to be aware of the workload and the various procedures that need to be followed to complete the task on time. This process will take time and it is possible that the team could be able to account for certain procedures that aren't crucial. It is also important to be aware that there may be delays in the production process when trying to comprehend the way Kanban processes work.

To ensure that the system functions optimally for the team, it is essential to know what strengths and weak points of each participant in the team. take note of the time spent by each team member to finish the task.

The manager should use this information to create a system and distribute the tasks among all members in the group.

Chapter 5: Project Management And Kanban

Project management is a term that is defined as the process of running an organization or following a checklist exactly to the letter. The concept of managing projects remains the same regardless the tasks it encompasses. It is essential to understand that the project manager is not the sole measure that is constantly changing within a business.

The majority of organizations that have utilized Kanban for project management have used Kanban as an instrument. to manage project management have a tendency to link the management of projects with visuals.

No one in the organization including the project manager is able to comprehend what's happening in a given process if or she is not able to have an illustration of the process. So project management is the process of arranging the steps to be accomplished before the objective of the project can be achieved.

What does Kanban assist?

With Kanban work, tasks can be divided and distributed among team members. Each team member will be expected to control their workload according to their needs and priorities.

Each member's work assignments can be easily reviewed and a sketch could be created to help understand the issues that a member might face when working on the task.

If there's an image of the procedures it is more clear the need for new employees to take on the work. It's also more clear when a worker needs to be directed.

The efficiency of team members and individuals can be assessed and measured by using the metrics displayed on the Kanban board. It is recommended to utilize digital systems as they calculate the calculations to reduce the amount of time required to involved.

Each member of the team learns what processes the organization is following and makes the role of each participant easy to grasp.

Common mistakes that Project Managers make

Project managers frequently make the mistake of micromanaging employees. They seek control and empowerment away from their employees , resulting in the feeling of a lack of responsibility. They may also feel they do not have to implement the process to reach the end goal.

Each team must ensure that it's self-organized. All members of the team must plan their work according to their ultimate objective. They shouldn't be looking at a step-by-step plan. In the event that a supervisor is micro-managing his team, he's inefficient and wasting his team's time.

What are the benefits of making use of online tools for project management?

If a manager of a project uses online tools, they will be able to determine the member who is in a hurry or unemployed and evaluate the progress of each procedure or task. They will help make life easier as they give the manager an idea of how work is reviewed, shared or discussed.

It is recommended to utilize tools that are online, such as Kanban because the manager will reduce the amount of time spent in meetings and communicating. Many companies use online project management because they are in large teams spread across various places.

What makes Kanban so efficient?

The simple nature of the tool is what makes it efficient. Managers and teams can write any backlog items in the section for tasks on the board and select the tasks to be prioritized in the present moment. They won't have to be concerned about missing some tasks due to task at hand.

There is also a feeling of accomplishment and satisfaction after the card is transferred from the list of tasks and onto the completed checklist on the wall. This is a great method to see the tasks that have been completed and those that remain in the process.

What is it that makes Kanban adaptable?

Kanban instruments can be employed across departments, and in a variety of ways as it's an easy tool. There aren't many processes that Kanban will not be useful. The majority of organizations have procedures that are possible to envision to make a busy day less stressful and more organised.

The manager as well as the other employees are able to plan their work and understand whether everything is in order with the help of this Kanban board. When the supervisor is feeling stressed all he or she must do is look to the Kanban board, and slow down.

What is the best way to use Visual Thinking help in planning?

The saying "A photograph will speak a thousand words" is the reason it is crucial to visualize the flow of work. It is more effective to visualize the tasks that need to be accomplished in the day written out on the board, rather than consider the tasks that need to be accomplished and then forgetting many of them during the course of.

It is thought that the majority of people are visual even when they do not intend to. Visual thinking can help create order and an order to the often chaotic and unorganized thoughts.

Are there any plans to move toward the use of visual project management generally?

This is because the majority of firms, regardless of size, recognize the value that planning and visual management can bring to an business. This is evident by looking at teams with smaller numbers. When teams that are smaller within an organization start to work efficiently, the company will seek out more teams to utilize the visual method of managing projects. As knowledge about Kanban is spread across the marketplace there are more and more who are looking to integrate the software into their businesses.

If your team is trying to improve efficiency, take a look at these points below:

Avoid multitasking as it can be an obstacle. There are occasions when you'll need to multitask , and it is logical to do it.

Take breaks as you'll feel more relaxed and be able to perform more efficiently.

Make sure you do the toughest tasks first. It is recommended to plan the day get more manageable by the hour.

* Turn off your phone or change it to silent while you work.

Do not schedule too many meetings.

• Prioritize tasks frequently and ensure that the sequence of work meet the requirements of the company.

Chapter 6: The Use Of Kanban Systems

Kanban is a very simple program that isn't just using whiteboards and writing down the duties of a group using various cards. In the first 2 chapters, it ought to have taught you the fact that Kanban systems go beyond they are. A business will gain by the use of Kanban systems when they follow the principles and procedures of the Kanban system.

The latest trends suggest that Kanban is being utilized across a variety of industries and sectors and is growing in popularity. Small start-ups, small agencies and even established companies are beginning to adopt Kanban methods.

Kanban within Software and IT

Kanban isn't a project management as well as a software development instrument and this is evident from the beginning. Kanban does not discuss the way software should be developed or offer an array of strategies to apply to manage a task. It doesn't discuss the way a process can be designed or how software should be put into place.

Thus, Kanban, unlike Scrum is a method which helps organizations and the teams within it to improve their work. Microsoft utilized Kanban in 2004 to help with its software development processes Since then, it has been utilized in various teams within the company.

The great thing about Kanban is that it is able to be applied to a variety of processes and methodologies. If an organization is employing agile methodologies such as XP, Scrum and others or traditional methods such as waterfall or iterative, Kanban can be applied to these methods to enhance processes, enhance the productivity and decrease the time to complete a task. This will allow an organization to maintain its commitment to providing goods and services that are of top quality.

Kanban as part of Software or Product Development

Many product development teams in the tech industry and development teams for software applications have employed Kanban to apply Agile as well as Lean principles. The Kanban method provides teams with the best rules and guidelines that assist them in visualizing their work and provide products and services with rapid speed.

They are also able to be constantly able to get feedback from their customers to aid them in improving their procedures. Teams are also in a position to promote their products to their clients faster.

The Kanban system has gone through changes over the past five years across a wide range of sectors, including the IT sector. Kanban is widely regarded as one of the best strategies to be employed by teams to control and improve services over time.

It is also a good idea to use the Kanban method also equips teams with the necessary methods and guidelines that help enhance Service Level Agreements (SLA) and reduce the risk involved in the process, cut down on the costs of delays and ensure that products are delivered in the appropriate time.

Kanban assists teams in delivery and customers to collaborate effectively, using concepts like of Services, Class of Services, 2-phase commit and deferred commitment to make sure that the proper procedures are implemented at the right time.

Companies have started to adopt Kanban following the introduction the Portfolio Kanban. Enterprise Services Planning and Upstream Kanban can be used to boost the performance of markets and to achieve greater agility.

Kanban as well as Enterprise Agility

The Kanban method assists an organization to improve the speed of delivery of goods and services over time. It does this by eliminating any obstacles or bottlenecks in the system, thereby increasing the efficiency of workflow and decreasing the amount of time required to finish the task. This helps teams provide constantly and get feedback from their customers within an extremely short period of time. This feedback can help the team to enhance the product or service and help the team be more agile.

The Kanban system implements the principles of the Agile Manifesto and aids teams provide the products and services demanded by their customers. The Kanban system is a combination of agile methodologies and methods to enhance processes and improve the efficiency of the company.

Kanban is an ideal solution for a lot of non-IT-based business processes as its roots are in manufacturing. If a company would like to become more efficient and flexible and produce high-quality products the company should make use of the Kanban method.

Medium and large-sized companies have been using 6-Sigma as well as Lean initiatives to improve their production processes for many years. But, Kanban systems enable every company, no matter its size or nature to boost every aspect of their business, including Procurement, HR, Marketing sales, etc.

Kanban is used to a variety of contexts for project management, including construction and engineering projects. Many organizations such as recruitment companies and staffing firms insurance companies, advertising agencies and many others use Kanban systems to cut down on the waste they produce and to simplify their processes, thereby increasing the quality and efficiency of their operations.

Chapter 7: What You Can Use Kanban To Improve The Complete Value Chain

Kanban systems are great for teams, groups or teams that are part of teams as they are a collaborative approach. This is the reason that boards are able to serve any department in business, as well as for the entire business.

With Kanban every department that is involved in the creation of services and goods can discover ways to optimize the process. This chapter outlines how each department within an organization can utilize Kanban to improve the efficiency of its work.

Human Resources and Onboarding

Internal processes, specifically HR processes, benefit from the utilization of Kanban boards as they allow you to arrange every task at the ready. A good example of such a process is hiring process. When a company announces open positions, a lot of things begin to happen.

It is likely that there will be a lot of applicants who are trying to contact the HR department regarding the job opening as well as the description of job.

• The HR department is required to collect all resumes and applications that are submitted in the mail to their office.

If the resume and application meets the required criteria, HR should contact the applicant and talk to the people responsible to arrange interviews and walk through the various steps involved in every hiring process.

This can be very overwhelming and overwhelming for an HR department, as there are numerous applicants to track as well. The HR department needs to know at what level each applicant is at. A Kanban board can help simplify the process and ensure that HR department is informed of the status of each applicant throughout the process.

There are occasions when a single board can't manage the amount of applicants. In these situations, Kanban project management software can be utilized. The program will not just keep the track of all the information on the board as well as the card that are on the table, can also track each applicant.

There are numerous internal processes that HR has to manage like onboarding. Because these processes are governed by processes that are easily handled by using the help of a Kanban board. For instance, a visual tool could be created to keep track of every new hire and check whether they've been properly educated. Kanban boards are great for those short-term projects that HR as well as the Learning and Development Teams must tackle all through the year.

Purchasing

Companies that depend on receiving deliveries can make use of Kanban boards and cards to track the items that are arriving and going. They also track the time when goods are brought in. Kanban boards can be used to keep track of when the products are arriving. Kanban board is made up of cards that permit every person within the department to be aware of the status of each shipment.

A Kanban board gives department's members with an summary of what's happening with the shipment which eliminates the need to go through the files each day to see how things are progressing.

This boosts efficiency and helps every member of your team understands when should anticipate the arrival of the product. If the shipment has already been delivered to the facility the board is able to identify the location where the items are removed and the manner in which they will be distributed.

Development Department

Every company that has a department for product development can reap the benefits of Kanban systems due to its collaboration capabilities. A Kanban system allows the completion of a project efficiently and quickly. Whatever the case, whether it's an ongoing project or web development team using an Kanban board can assist in the processes as the team can see how the process goes from beginning until the point of completion. Kanban can be used to accomplish a number of things but at the center of the Kanban system is a tool for managing projects. It is important to keep in mind that Kanban was introduced in Toyota to ease manufacture. Production, development assembly, manufacturing, as well as supply divisions of Toyota employed this method.

Sales

As a visualization and tracking instrument, Kanban boards help the sales department to keep on top of the sales progress. It is crucial that the department track each aspect involved in the sales starting from the initial lead until the conclusion of the sale. Additionally the board can also assist the department in keeping on top of any post-sale actions including issues and feedback.

Because all information is readily available in images, it will be easier and easy for departments to enhance the sales process in the future. This can assist businesses in working on their sales techniques and assist the team to understand the processes that work for their company as well as how to benefit from the results.

The other advantage of automation could be gained by using this Kanban software. For example, if leads on the board don't contact the department in search of more details Certain triggers could be utilized to send emails to the leads in order to keep them engaged with the process and then move these leads to the next column.

A Kanban board can help members of the sales team identify the leads who have progressed into the next stage in the sales process. The leads may be color-coded in accordance with their movement throughout the process. This provides the team with more details on the ways leads can be transformed into customers.

Marketing

The Kanban board can accomplish much in a marketing department as well as for general marketing. Each marketing strategy has a variety of techniques and it is difficult to track the various strategies. There are various departments in the department of marketing who often clash since they're not on the same level. For example, the marketing department might have to manage:

* SEO and Content Marketing

* Marketing via social media

* Video Marketing

* Email Marketing

* Print Adverts

• Connect with potential agencies

* Connect to media or channels

These are the essential things that the marketing department has to perform and it is crucial to work in tandem. The Kanban board is a way to organize all the initiatives that the marketing department must create into a more unified schedule of tasks. This helps the department to stay on the right course. The department could use an easy board to display the following:

* What tasks are included in the inventory list or backlog?

* What projects are currently running at the moment?

* When can these work be finished?

* What's the next tasks?

If the department wishes to make use of a more complex board, and also look at additional metrics, it could include additional columns to aid the department in understanding the following:

* What are the people doing for the department?

* What isn't doing for the department?

* What projects are falling in

* What areas should be changed?

Managers can pinpoint the various areas they will require more effort if they are in a position to visualize the work that must be done.

Customer Support

Kanban boards and systems can aid the customer support team handle customer questions or complaints swiftly. The team has to deal with clients and respond in a timely manner and help them. Customers might want to submit feedback, queries or complains at any moment.

If the department utilizes Kanban systems that allow for email, every request or telephone can be converted into a credit card, and the card could be put at the high point on the screen. If the team is able to see the card, it should be transferred to the next stage immediately.

This ensures that the process is going smoothly and everyone within the team knows the way a request from a customer is dealt with.

Helpdesk and IT

Support tickets can overwhelm and overwhelm the help desk. A Kanban board can arrange those tickets according to priority and then move them around the board. If you are using a Kanban software is utilized the tickets can be purchased based on priority of the order using the color codes. Members of the team that are skilled at managing these requests can select the tickets.

This also helps the help desk find more important issues and to determine what can be done to address these problems. If there are many individuals who are suffering from the same issue and the department is able to identify how to improve to enhance performance and reduce mistakes. Kanban is a system that can be used to identify problems. Kanban system can be used to detect problems.

Operations

Departments within business represent distinct components of the system and have to work together in order to reach an end objective. The operation team is responsible for holding all departments together, and is accountable to ensure the proper functioning of each department.

Because Kanban boards permit each department to know how they are doing with their work, they can assist them and their operations to make sure that they're on the same page as other departments.

Kanban boards are a fantastic method to link information flow or work across departments. They are able to discover the ways in which work from one department impacts one another.

Chapter 8: How Can Kanban Be Utilized In Back Offices?

It is essential to understand that Kaizen isn't an instrument to use with Kanban and actually the opposite. Kanban is a process or tool that is utilized by teams, individuals and companies to think more about their tasks and how they can function efficiently to accomplish the tasks that are in front of them. The Kanban method is intended to change the manager's brain. How do you make Kanban be utilized in the back office?

Imagine the very first task you're asked to complete in any self-organization manual. Each night or evening take a seat and write down the tasks you have to complete the next day. When you arrive at your workplace, pull your list out and begin the first task.

Make sure you don't take a leap into the third or second task without having completed the first. Make sure to only complete the second task once you have completed the first task. When your work is completed you can continue your day in the manner that you like.

This is an effective instrument as our minds are busy with various tasks and the various tasks we have to complete. Sometimes, we fail to finish things that are crucial due to the amount of work to think about. If we don't write prioritizing tasks then we'll start to delay and take off on other projects before the ones that we must take on right now.

Many times, people work on the first project they think of when they have too much to finish. If they run into a roadblock then they abandon this task and proceed to the next. This process continues until the individual has completed each of the tasks on his list, but hasn't accomplished any task. The person will find out that the entire morning has passed and not a single task has been accomplished.

Let's take a look at this example. Bob arrived at the office at 9 am and was working on four different tasks at once, without finishing every task. It's getting close to 5 PM and he is ready to go home, however there's nothing to report on the day. A new job was just announced and he would like to complete it but is unable to do so as he has four remaining tasks to finish.

He turns to Alice asking her to finish the task in the specified time. This results in a backlog, or a record of the tasks to be completed on Alice's desk, since she has a list of tasks to finish.

The Kanban system stores every new task on a single board and then assigns it to the team member when the team is completed with all tasks. This kind of work comes with the advantages listed below:

Each member of the group is assigned a task and finishes it completely. If the tasks don't require a lot of time, they may do three or two tasks to complete them in the specified time. In this way, they are able to concentrate on the task that are in front of them. If the backlog or inventory list is less, each individual will face fewer hurdles to tackle.

* If a group is assigned the next task, team members must consider the next

Is this project prioritised?

Why is this job need to be completed?

What's the advantage of this assignment?

Who is the person from whom this task will gain?

* This can help the team to think about whether the task is relevant and the people it impacts. This will allow the team to develop a habit of focusing doing tasks that bring value to the team.

The Kanban system allows every employee to contribute value to their job. The intention behind the system is to ensure that each employee performs at least one job that enhances the performance for the entire team. This is a standard to set for each person on the team. In many organizations, Kanban systems are used as agile methods. Teams write their workflows on a paper and observe how the various tasks are progressing throughout the week or day.

They can also see certain obstacles and determine ways to get around them. But, these systems will not take care to clear the backlog, or load on the desk of any team member. Also, it doesn't analyze how a specific task can be beneficial for the entire team.

The majority of managers favor flexible Kanban systems as they can control the process, and collaborate with a few colleagues. However, they do not understand the value in Kanban. Kanban system. They don't realize that

* Based on the actual demand of customers

* Reduce any backlogs in the hands

* Accelerate the flow of production

Kanban systems are not able to work with out lean administration. The team might be growing every day but when there is no tool that can assess the value of the team, then the changes achieved cannot be considered to be efficient. A Kanban system should assess whether the value of the product has been increased and how it has allowed a team member to effectively finish their work while utilizing a small amount of backlog.

The majority of organizations have been using lean methods for a long time without attempting to comprehend the function of these tools. Today, the trend is to experiment with the basic fundamentals of the Kanban systems and then learn the tools.

This is a major requirement since each team needs to be clear about what they're trying to understand, and then find the best way to implement the lessons acquired.

Chapter 9: In What Way Does Kanban Aid In Forecasting?

It's surprising that organizations still rely on the practices of shamans, clairvoyants and shamans to forecast the future. They believe that they can use big data to analyze the habits of their market and customers.

The reality is that companies depend on the development team to provide accurate estimates. Instead organizations can make use of Kanban boards to get estimates based on real information and data.

Myth about Estimates

It is crucial to keep in mind that predictability doesn't increase when you make an estimate. The ability of any system isn't affected by an estimate about the time it will take an entire team to finish the task. It is crucial to keep in mind that predictability is a characteristic for the whole system. If an organization does not recognize this it will not be able to accurately predict the behaviour in the systems.

The science behind prediction

It is straightforward to estimate how long it will take a person to complete a task , since there are some facts to support the estimation. It's a different matter to determine the amount of time it takes an application to get an item from its initial location to the final destination.

If someone asksyou "When do you expect it to be finished within?" they mean to inquire "When do you believe that the item will have reached stage of production?"

The members of the company knows the time, and often they speculate or even make up the exact time. There are times where an organization has been a fraud and has made a guess. It's risky to calculate an estimate and can be an unnecessary waste of time because the estimate was made using incorrect information. The company must make the effort to study data prior to making claims. But, they don't have the time or resources required to conduct this.

Information Theory

When an organization is making an estimate, there's little information available. It is recommended to create an estimate later when more information becomes accessible as the work progresses forward. This is the reason that it is suggested that businesses employ incremental and iterative methods of development and financing.

Mathematics

When two dependent variables have linear dependence upon one another, any changes that are triggered by other variables are influenced by the largest deviation calculated by the previous variables.

Queuing Theory

The lead-time increases as resources are utilized extensively in systems that have varying requirements.

Lean

The efficiency of flow is generally lower for the majority of organizations this means that lead-time can be affected by the environment. This means that lead-time is not dependent on the size or complexity of one feature within the project, nor is it specific to the individuals or their abilities.

Methods for Product Development

Traditional Project Management

Traditional project management employs the following constraints when making the estimation":

* Schedule

* Scope

* Budget

A budget is created and the timeline and scope are negotiated after the company has made plans and gathered an estimate.

Agile

Agile is not able to fulfill the promises made by conventional project management. There is a deadline for delivery set by the business but it is possible to define the extent of the work is ad-hoc. The scope is identified in broad terms however the specifics are developed in time.

Kanban

The Kanban software cannot provide a guarantee or promise a specific delivery date that is based on undetermined variables. If Kanban systems are adopted and the company agrees to have regular supply of top-quality products. Kanban provides a commitment to each type of service and provides an organization the chance to guarantee continuous delivery and transparency as well as constant improvement in quality and lead-time.

How do we accomplish this?

Develop your Kanban system

The first step is to create a well-designed and organized Kanban system that is properly structured and organized. You need to know the demand delivery, the limits that the Kanban system has, the various types of work, rules, categories of service and various other aspects. If an organization is looking to establish a system that is predictable that is predictable, it should identify the beginning and end of the process. Within these limits is which you can formulate your predictions.

Follow the guidelines and principles

The company must follow the six principles and four principles in Kanban. Kanban tool. This allows the delivery system to be more reliable over time.

Know Variability

The predictability of the system can be improved with Kanban tools to reduce any variations in the process. Lead Time Histograms, Lead Time Scatterplots, and Cumulative flow Diagrams are the most effective method of analyzing how the system is working and to identify the areas that require improvement. enhanced.

Variability is viewed as an inevitable evil and essential if an organization wants to develop new processes, however not to the extent that the process becomes unpredictable.

It is crucial to provide for a certain degree of variation in the beginning phases of the process, but not during the development phase that follows. A company cannot be plagued by frequent production errors and unreliable processes. It is also not possible to have excessive dependencies with other teams , and queues for every team that manage the tasks and processes being managed by staff members.

A few of the causes of variance include:

* Internal:

Irregular Flow

Size # of Work-Item

Class-of-Service Mix

The Work Items Type Mix

Rework

* External:

Requirements Ambiguity

Irregular Flow

Batch Size

Environment Available

Requests for Expedites

Complexity Scheduling Coordination Activities

• Other Market Factors

Humans

Forecasting

Depending on how the company is using Kanban depending on the way it is implemented, it should be able to forecast certain work tasks between three and four weeks after the implementation of Kanban. If the implementation works and efficient, then the company can forecast a number of large projects. Let's look at how this could be achieved.

Forecasting Single Features for Single Features

The majority of companies are inclined to move processes to development lines with the expectation that the process will be finished in the future. They don't realize how capable the individuals on the team of development. These companies do not realize that the system slows down when the pipeline gets clogged with a multitude of development tasks, leading to excessive stress and numerous mistakes.

If data is not obtained using a system that is predictable and the company is unable to stop the cycle. Customers and other stakeholders continue to request estimates once they discover that the company has stopped in delivering as promised. The development teams continue to make up stories and the work is pushed around between departments in the hopes that something will get completed. This leads to a slow and unstable system that causes the operations or business team put more work in the pipeline.

If an organization is using Kanban for at minimum three weeks and has enough information to build the lead-time histogram. This can help the company give a reliable forecast or prediction. It is frequently employed to make a decision on scheduling.

If an organization begins way too early in its forecasting process it could lose the opportunity to work better using the resources at its disposal. If it's done in the wrong time, there is an opportunity to incur costs for delay. As an organisation is responsible for making predictions, the following questions should be addressed prior to making predictions:

When is the best time to start?

Do we need to put this task on hold to allow us to focus on other projects?

Is it too late to get started on this task?

What are the risks of delaying the task?

Should other options be considered when the risk isn't worth it?

* If risk seems very high do other tasks need to be left unaffected?

What will it cost the company to put off the release of our product?

What will customers react?

Forecasting Multiple Features Using Multiple Features

Troy Magennis, has developed several tools to aid an organization in monitoring and forecast the evolution in software as well as other items. His website is full of information which can be utilized at no cost. Forecasting multiple features by using Monte Carlo Simulations. You can calculate the probability of each forecast by using the historical data.

Chapter 10: Adding And Setting Columns

Here are the best tips to follow for Adding as well as Setting Columns.

"good" list "Good" List

To determine if the list can be described as "good," see if you can cross off all items in the list below. If you're not able to check off one item, go through your plan to make corrections on your list for your group. You'll want your digital board to mark all items on this list:

Choose to go with an electronic format or a physical one according to the workflow that your company and the needs of your team.

* Make sure you have the minimum number of columns that you are able to have. Anything more than 7 is considered to be overkill.

All your tickets will apply to the current WORKFLOW However, they also embody the complete process.

All of Tickets are "high high level," meaning not every small task is taken into account and instead tells an entire narrative.

* POSTS IN YOUR BACKLOG contain direct links to tickets.

* NAME THE TICKETS ABOVE WITH CLEAR and SUCCINCT LABELS.

* You have clearly defined guidelines to "Definition Of To-Do" along with "Definition of Done" to which your team members refer to ensure that they meet your expectations prior to proceeding to the next step.

• Balance your workflow by establishing the limits of your team's capacity for WIP and plan to handle bottlenecks and "showstoppers." Maintain a list "well filled" so that team members have something to do next.

* Reject products that do not conform to standards. The term "not meeting standards" can be an indication of items of poor quality, or products that are too large or outputs that don't conform to the "definition of"done."

NOTE: AFFIDIDING A TEAM PARTNER to each task on the "DOING" LIBRARY OTHERWISE the task will be returned to "TO Do."

You have a system in place to verify that the "DONE" items are actually done.

How to Structure Your Board to help Your Team

The columns of your board depict how your project is completed starting from the initial creating phase. Each step of the process is represented and thought of as the "pipeline" for your project. Kanban is a Kanban method prefers blurring the distinction between "stage" and "state," offering columns that include"the "state" associated with the project like "doing," instead of calling it something like "analysis" as well as "testing."

In many situations, you'll know what's happening in the process dependent on the person working on it, therefore it's not necessary to break it down into more precise "stages." Certain teams prefer and appreciate a few additional "doing" choices. These additional columns reflect the requirements specific to the team or project. Some examples of additional "doing" columns are:

* "Ready" or "Next up " to be decided."

* "Ready for Analysis" or "Ready to Define"

* "Develop" or "Implement"

* "Integrate" or other dependencies coming from outside

* "Test"

* "Done" or "Complete"

In addition, adding these column spaces between "to complete" as well as "done" might be too much for small businesses or specific structures, whereas more complex and larger organizations are likely to appreciate the visually progression of every stage that the method goes through.

Notes on the additional Columns

"Ready for Analysis" (or "Ready to Define"

These distinctions are only relevant for specific actions that are part of a workflow that is closely related to each other and that requires to be performed before a different action is performed. In many situations it's a waste of a column. The role of analysis should be given it's own Board. This is because the majority of analysis happens prior to the creation of the project. The vast amounts of analysis stored in this column and cause delays that are irrelevant to the success or completion of the task. Instead, make these actions as separate as you can, if it is possible.

"Ready"

That means you need to come up with the "definition of ready." This defines the prerequisites that have to be met before the work is done. When an item is added to the list, it is arranged according to the importance level. Items with the highest priority rank at the top of the list, with those with the lowest priority are on the lowest. Every time an item gets transferred to this column, its importance level should be taken into consideration.

"Doing"

These columns that are focused on development, however, sometimes particularly if there is no "Ready to Define" column the actions that are listed in this column need analysis to be carried out. If you choose to delete this column, you must remove the "Ready for Analysis" column," include the one-time analysis requirements listed in this list.

"Test"

This is a different column that could wasted time, based on the field you work in. For instance, companies who are controlled by external organizations that need to evaluate your product prior to finalization or deployment gain by having this column. Other industries that conduct their own internal testing ought to look at removing this column as well as adding the actions to other tasks. The majority of "testing" occurs when a job is ready for the next stage to ensure that your product's "good" prior to transferring the task to another individual or the next stage within the process.

"Done" (or "Complete"

This is exactly what they mean when they say the job is complete. The "Definition of Complete" is essential to this point since it will ensure that your team members place items on the list if they are in line with your criteria for being "complete." In various different industries "done" means that an item is launched or is ready for release. It doesn't mean waiting for the event of an outside source or that the product is put in a hold. The definition can differ between businesses but it should not be used to conceal the work that is still to be completed.

"Integrate"

If your business finds that they have completed their work on the task , but is left to be patiently waiting for an external intervention think about adding an "Integrate" column. This is where the team will be able to place things in the event of waiting for the external condition. Sometimes, this "integrate" could occur during the development process or near the end of it and the location the location where the card is moved following "Integrate" is dependent on the workflow you have and your product.

"Relevance"

Although it is not a straight column "relevance" is a reference to what the team is currently working on or is planning to tackle in the near future or the next release. If it's part of the larger work, ensure that all actions associated with the section chosen are listed on your board and not anything else, so that it is free of the noise and volatility.

Transferring work from column to column

Kanban is system that utilizes visual representation to show how work is going when team members work on specific tasks. The workflow occurs when someone moves cards onto the Kanban board to ensure the production's equal status. If you transfer tickets in a systematic and systematic manner this helps the system function effectively. It speeds up the movement and allows for a seamless flow throughout the system. The Kanban approach aims to create an efficient and fluid system that improves efficiency and performance.

This concept of Kanban is about the management of work, not the employees. It helps a person be aware of the processes that are involved in completing a task item. This means that the person can better understand the system and identify any issues that arise within the process. Anyone who can quickly recognize and resolves a problem helps to increase the effectiveness in Kanban. Kanban system. A seamless workflow guarantees constant quality products and offering services that are beneficial to customers who are interested.

A worker shifts tasks from one column section based on the sequence of Kanban cards. For instance, puts four cards in the column called "backlog" and each one contains user stories ranging from 1 to 4. The person arranges them in decrepit order, where the Kanban card on top will be the first one to move and the fourth in the bottom row will be the last. They then move them in an order of urgency into the column 'to-do and from there, the team members are expected to be committed to a task when they are ready. After the limit of the current work of the particular system, the individual chooses the first card from the 'in process column.

If the work-in-progress limit is two, the person takes two playing cards out of the 'to-do to do' lane and puts them in the 'in-progress column. The team members agree to the projects in the column 'in progress' and continue to work to complete them. When the task is complete and the task is completed, the team member takes the Kanban card for the relevant task and puts it in the column marked 'completed. Then, he takes up a new Kanban card for a fresh task. The employee removes it from the 'to-do column and moves it into the 'in process in progress' track. The workflow continues until the task is complete.

Each cycle is a representation of the workflow process from beginning to end. The less time it takes to complete the process the more efficient the workflow will be as well as the better-performing Kanban is. Kanban process is. Thus, this concept aims to speed up the process by controlling the workflow and reducing delays by addressing issues quickly.

Kanban uses all three of the principles to maximize productivity within the shortest amount of time. It is essential to understand these principles and follow them. It is important to comprehend these fundamentals as well as the Kanban system's implementation. They assist an individual or business achieve massive and continuous achievement.

Kanban Tools

Kanban refers to the process to organize chaos and reveal workflows that facilitate problem solving and the flow of value within the working process. Kanban tools permit this sort of organization to be carried out in a structured way and aids in achieving efficiency. The features allow people to understand and manage their workflow , while also being aware of the challenges facing their company. The most important Kanban tools are the board and cards that each person creates according to his or requirements and use the board to improve productivity or flow.

Chapter 11: Beyond The Board

Kanban is more than just helping you manage your workflow . it can also assist you to design and create products or services.

Humans can comprehend more complex concepts faster when they are presented with information that is visual. The old saying is, "a picture is worth more than a thousand word phrases." It's hard to find people who are willing to question this notion.

Images can drive a message across more quickly than any other text, and also attract more interest. Managers of projects are aware of this, and use strategies such as Kanban boards to make their point. Kanban board to maximize the benefits of visualizing their workflows.

Before you use Kanban to design an item or service then you should be asking yourself these questions:

1. My principal customers and my main users?

2. What requirements will my service or product address meet? What is the value it can bring?

3. What are the most important features to satisfy the requirements I'm aiming to be able to

4. What is the quality of the product against other similar solutions? What makes it different?

5. How can I expect to earn money through this service? What are the possible sources of income?

6. Do I have the ability to create this? My company can create and market this service or product?

You'll have a clearer concept of what your goal is when you answer these questions. They can help you make more informed choices, particularly when you are planning to create an automated workflow that is based on the Value stream mapping.

Kanban vision boards can create greater impact when shared with your team since they provide transparency, and reminds each person on the team of their ultimate purpose.

There are several methods you can use prior to creating your entire workflow, as they can help you to see things clearly:

Value Map of Streams

If you're developing products or services the process will allow you to create a clear and precise representation of the steps involved in the process. In the field of software for instance the benefit the software development company can provide to its customers is the features included within the software.

Value stream maps allow you to visualize all the steps that are essential in your workflow , allowing you to provide worth throughout the process from beginning to completion. This allows you to see every single task an entire team is working on in a single glance.

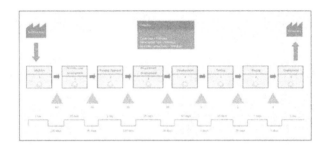

In the image above you can see a value stream mapping developed by a team of software developers. Each of the blue light boxes is a step through the development process. and the line of blue below includes the anticipated timeframes for each of these phases.

State Diagram

A state diagram can aid in understanding the reaction that an object exhibits when it is confronted by external stimuli. The diagram depicts the fluidity that an object goes through in its transition from state to state within the particular system.

When you are defining and making a diagram of a state it is important to think about all possible states for an object. If you are creating the development chart for a product it must cover every step it could be necessary for a product to be taken from its original condition to the final. For an online company, a product could be released on a specific date, an sold-out state and the added to shopping cart states, or saved to states on wish lists and the purchase state.

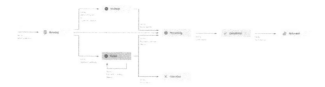

Workflow Diagram

This diagram gives a clear overview of the process in business. This workflow lets you visualise step-by-step the process from beginning to end using common forms and symbols. Workflows also serve to aid employees in understanding their role and how they impact the work process overall.

The speedy creation of workflow diagrams and simple way of expressing a procedure have been the primary reasons that it is employed across all sectors, from e-commerce to manufacturing to government agencies.

The diagram of workflow above was designed in the context of ServiceDesk Plus, a web-based, IT help desk software. The diagram shows the process users need to follow when they receive tickets or requests. It incorporates the various phases and workflow channels to provide greater visibility and clarity.

Once you've got an grasp of the procedure and each step that you need to model your Kanban board on the basis of the process.

The Kanban Board

Kanban boards are essential to the overall team. Kanban board is a key component of the entire team since it offers a unique way for team members to collaborate the board, share updates, and visualise their work. You can also post questions highlights, questions, or teammates in case they require assistance with crucial issues.

The components that comprise an Kanban board include:

Columns

The columns are used to organize the different phases of a task and their structure; these stages could be tasks in progress, current (To-Do) as well as work in Progress, and Completed.

Cards

These are the components which are moved between columns (stages). They assist in organizing and prioritize every job. They can be duplicated and removed to avoid redundant tasks.

Each card should best be a representation of a job, and should include the most details possible. Things like due date as well as priority level, project owner and estimates of amount of time that was spent on the task are highly appreciated.

Extending Your Kanban Board

While having a simple vision board is recommended but it is recommended to have a Kanban board may be more useful to include important information such as the name of the product, its design concepts, screenshots, and mock-ups. The data you gather from your market research is another piece of pertinent information that you could want to include in your vision board.

Turn Your To-Do List Into A Backlog

Although a backlog isn't a standard feature for Kanban, if you want to make it a part of your Kanban board, having one as part of your Scrum/Kanban hybrid could be a possibility. It's up to each team determine the right framework that fits their needs, and using the backlog in conjunction with your usual Kanban workflow will help you overcome long and unorganized to-do lists.

Because Kanban doesn't typically have an inventory of backlog items, members constantly add new issues to the top column. As the list grows it becomes difficult to organize and prioritize tasks. To avoid having to redefine your team, board or even your entire Kanban workflow, it is possible to divide your board in two screens: backlog for grooming of your backlog, and then the Kanban board for grooming your backlog. You can arrange cards according to the workflow and simultaneously logging the tasks you need to do in your backlog that will assist you in planning your future.

Visualizing Your Kanban Metrics

Each system requires data in order to make improvements which is why Kanban is no exception however, how do you track tangible values like improvements and workflow optimization? Since Kanban's focus is on communicating data in a visual manner There are numerous methods to gauge how effective an Kanban process has proved to be and the most popular of the bunch are diagrams.

Control Charts

The Kanban's sole readily accessible metric is the cycle time. The term "cycle" refers to the length of time required to complete a task or part of work through the entire procedure, from start to end. To aid you in understanding your cycles the control chart shows them in a timeframe using components like dots or bars. If your team is making constant improvement and is operating at its best it will show a downward trend of the individual cycles.

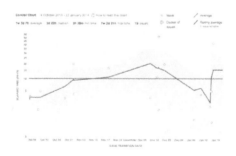

Control charts such as ones like the one shown above can assist teams in understanding how long it takes to

105

complete their tasks or to resolve problems. This particular example illustrates the time that is spent on problems and the time average, and the general date range, as it can help provide more clarity. It is important to always consider an average of rolling and then compare it to your overall average to see the work your team is doing.

Cumulative Flow Diagrams

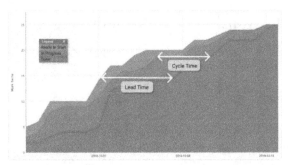

A CFD, also called CFD, is a type of CFD is a representation of the amount of work that is performed within an area graph. It depicts each step of the process over a specified time. Each color will represent every one of the actions that you have listed on the Kanban board. An overall flow diagram can aid in identifying trends, team performance and bottlenecks in a single glance.

This CFD is an excellent method to comprehend cycle times as well as lead times. While the process described is fairly straightforward but it can aid you in understanding how to monitor performance. the diagram can also be used for more complicated procedures, but make sure you review the size of the region that is each stage and then compare the performance with other stages.

Kanban Vision Board Examples

Here are a few Kanban board examples for various types of groups and sets of tasks:

Design

If the aim for the designers is to design user interfaces that provide a superior experience, then a suitable Kanban board might include the following steps:

* Prototype

* Implementation

* Review

* Done

Web Development

If the main goal of this group is to create enhancements, fix bugs and solve issues, its principal phases of work could include:

* To-Do

* Pause

* Reopened

* *

* Testing

* Live

Marketing

The aim of a marketing group is to grab attention and create interest through the marketing of a item or service. The workflow of a marketing team could include the following steps:

* Requirements

* Implementation

* Done

Sales

The sales team's job is to generate prospects and conclude deals. Its method could be illustrated by the following steps:

* Potential Customers

* Leads

* Contacted

* Pending Response

* Demo Arranged

* Final negotiations

* Win

* Lost

Customer Support

The Customer Support act of providing assistance and solutions to customers who have problems or have questions. The Kanban board may include these columns

* Announcements and Updates

* Tasks

* Bugs

* Quick Fix

* Service Requests

* Feature Requests and Suggestions

* KPIs and reports

* Done

Employment

The purpose of this group is to choose the most qualified candidates and convince them to be a part of an company. It could be that they follow the following process:

* Positions Available

* Checking Resume

* Rejected Resume

* Invitation to an interview

* Final Interview

"Make an Offer"

* Positions Closed

New Hire Onboarding

The primary purpose of this workflow is to facilitate the transition process for new team members to ensure they are at ease with the interpersonal and performance aspects they'll be confronted with. The workflow for onboarding new hires may include:

* Preparation

* Getting Started-First Week

* Getting Started-First Month

* Quarterly Review

* 6-Month Review

* Yearly Review

Welcome to The Team

This workflow is designed to give new members access to corporate and social information such as benefits, activities and policies, among others. A Kanban board to facilitate this process could include:

General Information

* Positions Open

• New Team Members

* Team Rules

* How Do I Place an Order?

* Team Events

* Corporate Discounts

• Social Media Page

It's more than just A Board

While the vision board can be an effective tool however, it's merely intended to be a tool to achieve a goal. It helps to facilitate the discussion of what the team would like to accomplish and makes sure that all those involved are involved.

It is recommended that you take only as much time as you can on defining what the workflow overall should appear instead of beginning to collect feedback, while identifying performance trends and bottlenecks. After that, you can continue to study and adjust.

Chapter 12: Implementation The Program / Start

Implementing Kanban

Create the Kanban board to enable you to see, in a glance, the progress of the task that is being completed. Plan your workflow as it is currently - maybe not the ideal method, or the one you want. Every step of your workflow gives Kanban cards after you've visualized your workflow and built your Kanban board. Then, you can begin adding tasks to it. Visualize each task by putting an image on Kanban. Kanban board. Each task should be a signpost to something that has to be completed and that is worthwhile to do. Every job should be given an official title that everyone recognizes and understands. If there is no clear knowledge of how the project operates and what exactly it is similar to, it's often difficult to have a discussion about how to improve the project. The individuals who do the work are the ones who evaluate the results. Imagine how the work should be carried out. There is a limit to the amount of tasks you can be working on, and you must be able to complete them efficiently. This limit is typically smaller than you think. Working-in-progress (WIP) limits can be used to highlight the bottlenecks that a group's process faces they may not even realize exist. It's essential to gain approval from the rest of the group prior to establishing WIP limits in order to ensure that everyone understands the advantages of limiting the WIP before you make this decision. It's difficult to comprehend the right amount of work prior to the time you start but you'll need to begin at some point, so you should start by making the best guess. When you've agreed to a WIP limit for each phase (column) within the Kanban system, draw the WIPs on the Kanban board.

After you've created the Kanban board, everyone is in a position to see the current status of what's being done and who's doing the work and where in the workflow every task is. Each day, check your Kanban board in order to understand where the work is at and work on getting rid of the blocks to move forward. Always begin your discussion on the right-hand part of your board. This means that you're always thinking about the client's shipping needs.

Keep an everyday Scrum

Nowadays, everyone is aware of the boundaries they have to adhere to in order to stay within the limits. When someone is ready to do tasks, they examine the plan to determine the tasks to be completed and then move the task to a column that represents another stage of the procedure. It becomes their job until they finish that part of the process and someone else moves it to a different column. In practice, this typically means that your team ceases to begin the new job and then finishing the work within your Kanban system.

Manage Your Partners

Your employees must determine the efficacy in the Kanban system by keeping track of the flow (guide) as well as time, throughput, and other indicators. The metrics can be easily collected by using tools such as spreadsheets. It's essential to track data from the beginning so that every change can be evaluated to determine whether they have an effect that is positive on the flow. It is essential to have the smoothest and fastest stream. This means we need to ensure that the Kanban process is producing results quickly and with a regularity.

Gather Metrics

Think about how many tasks you could complete in a week, rush, or other period of time. We want to understand the differences in the speed of work so we can compare the tasks completed from one time-frame to the next in the hope of being regular. The data that is collected will give you with the data needed to adjust your working methods and evaluate the impact of these adjustments in stream. Kanban gives teams flexibility in their preparation and predictable outcomes, as well as greater visibility and more focused the delivery. If your client is within or external to the business is going to seek out the benefits of this method. If your client is in exactly the same location and is in the exact same position, then they'll have the capability to observe what the work is envisioned. If not, you could decide to share this information via email in a particular way. Once your process is working and you are able to see that you must talk with your customer more often about the tasks they'd like to see you do.

The data you collect must justify any changes to working procedures your employees might wish to make. The data you gather can be incorporated into the existing coverage you're putting in place. The power of Kanban is that it allows for anyone to spot the issue that is forming. The early recognition of the issue means that an answer is possible until you've accumulated a large amount of work that's not yet completed. If there's no obstructions, or ineffective or unneeded steps, they'll eventually come to the surface on your own Kanban board.

Increase collaboration Enhance collaboration Kanban. As your understanding of the job and manner in which it must be done improves and improves, it should be evident onto the Kanban display, along with your process policies, and your metrics collection. Every job is different as should the plan and execution in Kanban and continuous enhancement.

The Reasons to Use Kanban to manage your job? Management

You manage groups, contracts budgets plans, groups, and the risk that could affect you. You need to keep your managers and stakeholders happy You probably also oversee subcontractors too.

With heavy-weight process-oriented PMMs the emphasis is more on meeting the requirements specification rather than providing the highest quality and importance needed. The focus shifted to the team as the PM got eliminated out of the equation for job satisfaction at the very least. Scrum works with a predetermined list of characters, events and artifacts. an entirely new language were established.

However, Scrum hasn't become the answer to all issues in the workplace.

The reality is that the prerequisites for a successful Scrum don't always meet. Agile teams work in groups that are not agile, and they aren't aware of the fundamentals of Scrum and what it means regarding team behaviour. Scrum's Product Owner position is not always fulfilled in a timely manner and not all people appreciate openness. Certain people may be at a loss when they see everybody else is doing, and also from the open discussions about the progress and efficiency.

Additionally, Scrum is challenging to implement in conjunction that have a strong command and control culture. It's also a huge issue to expand Scrum to the app and portfolio level. Although Scrum is great for smaller tasks however it's not suitable at the portfolio or app level.

Both can solve some of the issues however, they may not solve all.

The way Kanban helps make the lives of Job Supervisors Less Complex

When I first came across Kanban I knew that it might provide the solution I was looking for. Kanban is as much a mental model as it is a strategy.

In depth analysis of the Kanban process is not the subject of this article. But, there is one thing that is clear from the perspective of a PM With Kanban the team is always doing the most important tasks first, and as quickly as is possible. This pattern is evident through the five fundamental aspects of Kanban:

* Visualize your job

* Restrict work in progress

* Manage flow

* Make policy clear

* Create feedback loops

With Kanban you can manage various types of work on one board, while focusing on stream and completion of work.

This is precisely what jobs really need, since it keeps jobs from turning 90%, 95 percent, or ninety-eight percent complete over the course of time. Additionally you get rid of your actions lists, and Excel sheets.

Kanban captures and measures the various types of work for example:

* Technical work

* Follow deadlines

* Subcontractors' work

* Unique work

When it is feasible to limit work in progress (WIP) to the work we have in our possession however, it is not applicable to the work of subcontractors. In addition, it's important to quantify and visualize work that, at one point was not visible.

It is, in fact feasible to run this way across all kinds of organisations, provided that you're allowed to be able to. This is what's involved in managing projects using Kanban:

• How simple job reporting can still help them stay in the management

How simple metrics can reveal these patterns, unfavorable deviations, and the possibility of advancement

• How visualization helps them understand the actual status of their job at any particular moment

* Having a discussion with your team about the best way to use Kanban (no, Kanban isn't just a big task board)

* Accepting the functional principle "Stop starting -- start to complete"

Kanban Can Kanban Help You With All Your Work-related Problems?

The answer is no, Kanban is not likely to resolve all of your problems. The solution to poor job performance prior to is to add more managerial techniques, bureaucracy and procedures and sophistication to the PMMs which makes their use more difficult. Additionally, businesses that perform difficult jobs are unable to understand their role within the work environment.

The future will have options or issues that no method can solve. But for myself personally Kanban is making my working life as a project manager easier and has produced higher rates of success on my own projects.

Utilizing Kanban boards offers a myriad of advantages; one of the biggest can be the possibility of keeping everyone in sync. Kanban boards are able to answer a variety of the team's questions and concerns.

Kanban boards can help you see your tasks for efficient management. It's similar to making a list of tasks on the back of a post-it note. The visual aspect of the board allows teams collaborate more effectively and take on more work when they're able. This will boost your team's efficiency.

It's common for businesses that plan their needs for IT as well as software development companies to employ Kanban. There are many kinds of organizations that could and have used Kanban. They view it as a method to improve efficiency.

Another advantage that comes with Kanban board is the versatility. Since all you require for the basic requirements of Kanban cards is whiteboards and sticky notes, everyone can design or modify the board with just the basics of.

Kanban Boards are Easy to use

Kanban boards may be constructed quickly and are constructed in a most efficient way for every project. Every card is constructed in a matter of minutes and placed on the board where you require for it.

In any undertaking, people will spend lots of time in meetings and they're the method by which people meet to collaborate. Meetings aren't always easy as well as too much meetings can put an impediment to productivity. Kanban assists in limiting meetings to the time when they are essential. Kanban boards can assist in communicating changes since Kanban boards encourage transparency. The team is able to see the activities happening in real-time.

These could include status updates or other modifications so that the team as well as Project Managers are able to focus on completing the task.

Kanban boards are extremely effective in helping teams reduce anxiety, as they reduce work load and allowing teams be able to work as they wish during the process of project development.

You may be wondering why projects are slow to finish and why projects tend to take a long time to complete even though you feel like your team is working hard. Kanban boards are incredible because they allow your team identify the areas where there could be a blockage in the workflow or work that is not that needs to be completed.

We are all aware that people are more productive when they are able to focus on one or two tasks at one time. The more tasks people are required to complete and accomplish, the more you'll notice a decrease in the quality. Kanban boards could have a lot of items in their backlog However, everyone is focused on only one thing at a given time.

The ability to discharge at any time

When using Scrum and XP you are not able to launch during the middle of an the process. In Kanban you can release at any time. Once the story for the customer is completed, you may start it. It's not easy to establish development processes in such a way. It's essential to have "branch-by-feature" resources control, combine often and integrate, as well as test often. This is something that's worth fighting for.

The scrum is where it is not possible to make up stories during the Sprint (generally). It's not an easy process and often the development team is resistant to re-inventing a story from the backlog of sprints. The new story may have to be discussed within a hurry however, some details could be missed, and as a result major rework is likely to be needed. In general, sprints or iterations must be stopped.

In Kanban If you receive an urgent request to complete or a critical user story You can simply put it at the at the top of the queue.

There is no demand for iterations.

Why should you consider repetitions? In the beginning, iterations can help show that there is a problem in the development process speedily. Additionally, they establish an established routine for the job and establish rituals. At some point in your job, I'm assuming that you will no longer need them.

Our backlog is not as clear today and our strategies are constantly changing. We are able to be well, but iterations aren't necessary no more. They're useless. Instead, we've got shorter just-in time meetings with people within the team before starting the development of each user's story. It's a great idea.

Some people don't get the rhythm, however I believe it's more about dependency. Kanban creates a more complex rhythms and it may take some time for the development personnel to grasp the concept. However, a steady rhythm could be the only reason for sustaining iterations in the final stage of design, in my opinion.

No need for quotes

What is the reason you require quotes? In iterative development, you'll need them to predict how many stories you'll be able to tell during the next iteration. It is possible to predict an exact date for launch in relation to the backlog you've planned and the speed at which iterations are conducted. Another reason could be that the PO wants to know the size of the consumer story is. If it's big enough, PO might look at the possibility of moving it to a different release. If it's small, PO may choose to put it in a different version.

It's evident that iterative development is not even possible without quotation marks. If you decide to stop the iterations, then there's not an issue. Does the Product Owner have to live without quotes? We do not evaluate consumer stories. Why is that? As a PO, I personally an officer, do not wish to use them and don't use variations. What I would like to know is a rough estimate of what is normal, large and even really big.

Perfect Visualization of Flow

Kanban board offers a transparent view of the current work being planned. It shows the flow of work and facilitates quick scheduling and tracking. It's an incredible tool.

Kanban is becoming a well-known and effective tool. It's a great tool for specific software tasks of all kind. However, there's the risk of misunderstandings that lead to Kanban adoption.

It's incredibly easy to understand that you can't separate user-generated reports and therefore iterations have to be left-handed. The most straightforward solution isn't necessarily the best option. There's a reason why you are unable to divide stories. Most likely you don't know how to properly divide stories. It's a bit difficult at beginning and requires imagination.

In addition, as per Queuing Theory, it's much better to create small stories of roughly equal dimensions. Smaller groups with similar dimensions increase the flow (and you'll have a more accurate and reliable cycle time). In Kanban however, you'll need to break stories down into smaller pieces and try to create equivalent dimensions.

There are numerous reasons to use mini-waterfall strategy including speed-hunger, huge stories manual testing, inadequate descriptions of stories and so on. There is a chance that you are using an inaccurate length of iteration. For instance when you have a large job, one week of iteration could be considered an overhead that comes with a significant cost for trade. Therefore, a one-month timeframe could be preferable in this scenario. It is important to look into the real motives and the reasons behind the problem.

Chapter 13: Environment And Teamwork

How to Improve Your Workplace and Develop Your Employees

Employees are a major contributor to the growth or decline of your business dependent on the way you manage your employees. Through training, motivation and involving your employees on a daily basis and you can ensure a brighter future for your company.

In the event that everyone is trained within the company, all employees will eventually play a role to the eliminating waste. It won't take place over night but you will see improvements every day, small by little.

* MANAGERS TRAINING

Middle management and senior management should be taught about the importance of knowing how things are conducted from frontlines. They must have firsthand knowledge of what's taking place at work. They must learn the facts for themselves rather than relying on reports presented at boardroom meetings.

Engaging with front line employees and their tasks is a must for the managers and supervisors to ensure they have a clear understanding of the issue. Through this they can think of and propose solutions to these issues in the shortest time possible. Also, it is through them working in the workplace as well as with employees that they can identify the factors that could be contributing to issues that might develop later on.

If you're a supervisor or manager in your role, being able identify problems in the process will reduce time, energy, and resources that would have been used to complete the process if these bottlenecks not been identified earlier. The time, energy, and resources saved can be used to improve other processes.

* FIXING SKILL MISMATCH TO SKILL

Because you can visualise the whole and each specific aspect of the process it is possible to identify obstacles that could hinder working process. Sometimes, the bottlenecks are caused by employees. This could be due to their inability to be properly aware to their duties or understanding how equipment functions and the incompatibility between their capabilities and the work assigned to them, absence and other.

One of the issues that could arise during this process is the lack of understanding of certain workers of the older generation who have access to the latest technology in the modern world. The assignment of them to tasks they are able to handle in the present is the most effective way to proceed. In addition you must train your employees to use the latest technology that is being introduced to your business.

The ability to teach them new things can improve their efficiency and abilities. The older generation of employees put at their best with total commitment to your business. By helping them to improve their work by making them more efficient increases their confidence and lets them know that you trust them and will surely improve your sales in the future.

* QUALITY CIRCULES

It is a group comprising people working on similar projects. The groups are created to discuss issues that arise from workflow, particularly quality concerns and to come up with strategies for improvement. They're typically small and directed by an experienced mentor. In the ideal scenario, they receive instruction in methods for solving problems, such as brainstorming, cause-and-effect diagrams and so on. They'll then provide their findings and suggestions to management . Once they are given approval, the teams will handle the implementation.

* KAIZEN SPIRIT

Kaizen is based on the notion that continuous improvement is a regular element of the job and not something you can only do in the limited time after you've completed everything else. Individually-generated suggestions and quality circles can assist in improving the efficiency of the job during the course of the day. Instruct employees to keep an eye out for ways that the system could be improved. Accept suggestions. One method how this could be accomplished is through the Use of Andon.

Andon is an Japanese word utilized in the lean production method it is a method of notifying employees, management and the maintenance of issues regarding quality or the process. It could be done by manually activation of a worker using a button or pull cord. It can also be automated by the equipment used in production itself. The process will stop until the issue is fixed.

The majority of businesses use computers or software to perform work. Although these are tools but automated work still requires specific human judgement in order to get things done in the correct way. This means that many machines can't be left to perform the job given as there's a risk that something might be wrong if there is no one who supervises them.

The continuous transfer of human judgement to the system, so that the system is able to monitor on its own without call outs of humans whenever it senses that something is not right is referred to as automation. It is crucial because it is a way to separate the human from machines and keeps humans from doing the tasks assigned to machines. It encourages everyone within the organization to seek ways to create better and lighter machines that will cost less capital expense.

Lean thinking in the workplace implies encouraging everyone to be can think together and that nobody should have a challenge to deal with on their own. The process of Andon lets knowledgeable employees stop creation of products if the defect is found and request assistance in resolving the issue. Andon is a teacher of employees to think lean by highlighting the obstacles on-the-spot in the pursuit of being free of defects in all phases of the the entire process always.

* Use the PDCA Technology

It is an acronym for Plan-Do, Check-Act, which is the four-step process of the iterative approach to management that was developed through William Edwards Deming, an American statistician who worked in Japan to teach the leaders of renowned businesses. It is a good idea to use it to solve problems.

Let's say you're having an issue that is affecting the speed at which your customer support team is able to handle complaints. Many customers are not happy with the length of time it takes to receive a response. Here's how to use PDCA to address the issue.

1. Plan. At this point, you make the effort to create the plan of action for what has to be completed. If the project is large the likelihood is that making a plan on your own will take much work and energy. It is possible to make the task easier to manage by breaking it down into smaller steps or breaking it down into smaller segments. At this point you'll need to pinpoint the main problems that require to be solved and the resources you currently have and the ones you'll need to purchase. If it's impossible to obtain the resources you require, you'll must develop a strategy regarding what you could do using the resources you have.

It is also important to establish an "win requirement." This is an objective that will enable you to determine whether your strategy is effective. For instance, your win condition could be "Only three percent of customers who complaints mention slow response rates as a reason for the issue" or "10 hours of work have been devoted to customer service".

Read the plan together with the entire group at least a couple of times before beginning the next step.

2. Do. At this point you will apply all you learned in the previous step. It is recommended to execute your plan in a small scale as unanticipated problems might be encountered.

In this case, for instance If the issue is the way in the customer's concerns are addressed in the planning phase you've concluded that the reason is because there's no team that is geared to this job, think about the possibility of training some workers to give more time to customer service instead of creating an entire department for customer service. Be aware that this can have repercussions across your entire business system since the time that is allocated to assistance to customers will have been eliminated from other tasks.

3. Check. This is an essential step in the PDCA procedure. Examine how the plan was implemented and determine if it was successful. Examine if the win-condition was achieved. Have the amount of people complaining about the lack of response go down? If yes however, you must verify if the amount of complaints from customers is the same as prior numbers. Did employees devote more time to customer service? Perhaps there was a decrease in complaints since the company had been slow in the initial place. Perhaps, employees did spend more time they put into customer service but not enough.

It is important to determine whether the issue was solved and if not you must analyze it to determine the root of the issue.

Act. If all seems to have gone according to plan and you're ready to implement your original plan on a bigger scale.

If you come across an issue regarding customer service you could go through these steps over to make minor changes.

It is believed that the PDCA cycle is an extremely effective tool to fix issues at any level within your company. Because it's an iterative process, your team will continue searching for and testing solutions as time passes and make minor changes. Be aware that this method takes time, therefore it's not a good idea to employ this approach to tackle the urgency of a problem.

* SMED

SMED initially named Single Minute Exchange of Die or the change of tools in less than 10 minutes, is a key method of lean thinking that specifically focuses on the flexibility.

Flexibility means the ability to swiftly move from one task to another. SMED is a lean training program that teaches striving to increase flexibility continuously until there is a continuous flow of work in the correct time that can react to the needs of immediate customers.

* STANDARDIZED work

Standardized work is an illustration graphically depicting the seamless flow, having zero or one working-in-progress, and a clear place for everything, even the steps. This is a great way to implement lean thinking-based management techniques. Lean thinking is about aiming to the fastest and most efficient process in every project by finding and solving issues one at a time which leads to the improvement of the workflow as well as the independence of employees.

If you have a standard work process that you provide employees with guidelines regarding how to do their work. But, as time passes the guidelines need been revised in light of what is required in Agile development. There are changes and improvement are implemented, and you have to ensure that everyone in the team is in agreement. If a team member is struggling and a bottleneck forms in the sections of the board they are working on, everyone else on the team is affected.

Here are some suggestions for making sure that standard work is carried out when working in An Agile environment.

1. Prioritize the procedures which directly enhance your business - those which affect production quality and speed. Of obviously, there are other items that are not of any value, but they must be followed like safety standards.

2. Create best practices to complete tasks. This can be accomplished through seminars with the team. The process should be as transparent as is possible, and the suggested processing duration for each step must be clearly stated. Consider incorporating checkpoints that employees

could make use of to identify the deviations. Examine to determine if the guidelines or directions are appropriate.

3. Show the process and task visually. This is the point where Kanban boards help. Managers and employees must be able to clearly communicate what must be done and whether the task is being completed in a proper manner. This will help in determining the areas that could be improved.

4. Each team member should be trained according to the guidelines. Be sure everyone is aware that they must adhere to the rules. Be evident that you'd like everyone to signal immediately when a problem occurs.

5. Leaders of teams should be aware of the fact that they are expected to start improvement projects. It's better to standardize certain elements of the manager's job and duties. Managers should be aware of the things that could

slow the flow of work for an upcoming day.

6. Use an improvement procedure. The board may use indicators to show that the task was done in accordance with the standard. For instance green cards could be used to signal that the task was completed smoothly while red cards could be a sign of some deviation. This will allow you to detect problems that occur frequently. Regularly hold meetings to talk about issues and suggestions for improvement.

Standardized work enables employees to recognize difficult aspects of quality so that they are able to visualize what is important to the client and be able to discern the right and wrong at any point and confidently move from one step to the next. It helps employees develop lean thinking by allowing them to see the source of each obstacle to smooth work , and also by highlighting areas that are suitable for Kaizen.

Think about educating your employees on lean thinking in order to enhance their capabilities and skills that can then be utilized to find and address issues within the company. If you combine it together with the Agile manifesto and the Agile Manifesto, overall performance will improve. If Scrum Framework is integrated into your system Scrum Framework is integrated into your application allows for work to be made easier for all members of the scrum teams who have their particular duties and contributions to the overall success in the development.

Chapter 14: The Benefits Of Kanban

The idea of managing the development process of the product is not something that is brand new. The underlying philosophy behind Kanban is well-known since its inception during the 1940s.

The principles behind management of product development and creation of knowledge on as well. Taiichi Ohno's Toyota Production System (TPS) and Lean Manufacturing originate from Kanban.

If a business decides to control work rather than managing people, it can help to create a more balanced and humane, creative, and a synergistic environment for people to reach their maximum potential. This is a straightforward approach to implement an organizational change. This method isn't intrusive and has been adopted by a variety of new companies because it begins with managing the work being done within the company.

Let's look at ways Kanban can benefit your company.

Business Values are the first priority

David Anderson, who is known as one of the original pioneers in the field of Kanban technique, declared that Kanban is not just a technique or a method to control the progress of projects but it is also a framework for decision-making which is extremely efficient. The Kanban system was developed to allow an organization to take actions which are specifically based on financial goals.

As organizations operate in a competitive environment They must prioritize, plan and execute tasks as swiftly that they possibly can. They should also be striving to ensure that their work is error-free , to stay in front of their rivals.

Don Reinertsen has said that any business should be able to quantify the expense of putting off an item. In addition, the business may also look at what the price of installing one feature over another might be, as the feature being implemented may represent the feature that makes the company apart from other companies within the same field. The company must also consider the time it takes to use one particular feature over the other.

Enhances Visibility

The majority of the work done by the organization takes place in the shadows and it is vital to present that work to the leaders of the organization as well as its clients. This is one of the main features of Kanban. Kanban boards can serve as information radiators to observe the development of a process as well as any obstacles or bottlenecks that may be during the process in a single glance.

The information is accessible not only to employees of the company, but it is also available to all interested parties, customers, or observers. This allows for the sharing of boundaryless information within the company. The limitations of the software aid in understanding the way in which work is assessed and the way it is prioritized.

Reduces Context-Switching

The amount of work tasks being handled by employees is counted within the Kanban system, and is limited by the WIP limit. This system is focused to help employees to complete tasks that are both important and of high value, which improves its value to your company.

Kanban helps teams avoid overloading their members by imposing personal WIP limit. Every member is allowed to begin another project when the member has completed the task they've committed to completing. The Kanban list follows the following principle : 'Stop beginning then start ending. This method has helped teams to focus and deliver their downstream activities successfully.

Improves Collaboration

In many companies, departments are separated from each other. There are instances where there are conflicts that are fought between teams for software delivery and the product management. With Kanban the teams are part of the value system for development.

Kanban is pull-based system that helps to create synergies. It also helps break down the barriers between departments and specializations. This can results in cross-departmental collaboration. The process of transferring work items to Kanban boards gives teams the chance to share their knowledge as well as collaborate and interact with each other.

Reduces the wasteful activities

Many project managers are focused on timelines, not process queues as the former is deeply rooted into their minds. They employ Gantt charts as well as other documents to assess the timelines for each participant in the team. The thing that most project managers fail to realize is that they need to embrace the possibility of uncertainty, not just the granular planning. Many project managers try to create activities that add to the duration of the project and the risk involved which in turn increases the queue of process.

The Kanban board imposes WIP restrictions which make it an underlying pull-based system that lets a team or organization to have a consistent supply of ideas with high-quality that can be implemented in the appropriate timeframe. It also assists in eliminating any work considered inefficient, thus reducing the amount of queues to be built.

Some of the activities upstream like workshops, gathering requirements and business cases happen in the event that they are required. That means the manager of the project has to take quick decisions. Kanban lets managers to manage multiple projects pertinent to validate and prioritize ideas across the board.

The company introduces true sustainability

Kanban methods help teams organize their work at a steady, sustainable and fluid pace. This helps reduce stress, frustration and lack of dedication, thus raising the likelihood of employee turnover. Limits on WIP can help manage the pace of the process in a dynamic manner, which encourages creativity.

The teams will not commit to a particular process at first and then breach that commitment later in the future. Teams will be able to tackle and develop issues in a variety of ways to develop solutions that do not have as many quality problems.

Enhances Quality

For most professionals, the initial quality is a crucial aspect of a successful project. The impact of errors can be devastating to the productivity of the team. They also aid the team in tackling any quality issues at the beginning that can improve productivity. There are many tasks, such as collaborative analysis, user documentation, that help to create quality software.

The policies of Kanban software are designed to strengthen standards for professional conduct which are shared across various boards, such as project managers and software products, product manager, corporate stakeholders, and customers. The software clearly defines the policies at each stage in the procedure.

Enhances Morale

The Kanban software lets many agile and traditional teams to shift away from command and control methods. Each member of the team is required to create their schedules according to the workflow managed by the system rather than their bosses. Each member of the team must perform their work at a fast steady, consistent and sustainable rate, which helps induce a healthy stress known as the eustress. This allows team participants to see the fruits of their efforts. Because Kanban is able to manage the work of each person in the group, they will not feel overwhelmed, and they are able to focus on their work and develop their talents and creative abilities.

Instills Kaizen Culture

Kaizen board Kaizen board is a system that provides team members and managers with queues that can be controlled and need fewer buffers. Pull-based systems help identify any delays, process inefficiencies obstructions, problems with delivery and communication, agency issues and synergies, as well as absence of clarity.

These queues bring attention to problems and assist members and teams identify ways to address the issue prior to development slowing down. With these lines, Kanban introduces Kaizen, which is a term used to describe a process where teams learn to continually improve its approach to dealing issues all over the board.

Introduces Predictability

Kanban was believed to be an Information reservoir during the time of its development. By using lists and queues that are displayed on the board, Kanban can help a manager anticipate the most common issues based upon actual or historical information. This decreases the chance of any guesswork that a project manager might make. When changes are implemented using actual data, time to deliver and lead times are improved with time.

There are many advantages of the use of the Kanban system. A company will be able to cut costs and improve efficiency by using this Kanban system. There are few things to bear in mind prior to installing the system.

The team must keep track of the work load and the various procedures that need to be followed in order to meet deadlines. This process will take time and there is the possibility that the team will have to take into account some procedures that aren't crucial. It is important to be aware that there may be delays in production when trying learn the ways in which Kanban processes work.

To ensure that the system functions best for the team , it is essential to know what strengths and weak points of each member of the team. You should take note of the time spent by each team member to finish the task.

The manager should use this information to create a system and split the process among all members in the group.

Project Management and Kanban

Project management is a term which can be described as the process of running an organization or adhering to a checklist exactly to the letter. The concept of managing projects remains the same no matter the tasks it encompasses. It is crucial to keep in mind that the project manager is not the sole scale that can change within a business.

The majority of organizations that have utilized Kanban for project management have used Kanban as an instrument. to manage managing projects will link the management of projects with visuals.

Everyone in the company even the project manager is able to comprehend what's happening in a given process if or she doesn't have visual representations of the process. So the term "project management" is the process of arranging the steps that must be completed before the end goal of the project can be achieved.

What is Kanban and how can Kanban aid?

* With Kanban the work can be divided and assigned among team members. Each team member must control their workload according to their preferences and priorities.

Each member's work assignments are easily reviewed and a sketch may be created to help understand the issues that a person may encounter when working on the project.

In the event that there's an image recording of the process it is simpler to comprehend the need for new employees to take on the work. It's also more clear the need for a person to be directed.

Efficiency of teams and individuals can be measured and identified through the metrics shown in the Kanban board. It is best to utilize digital systems because they perform the calculations for you , thus reducing the amount of effort required to performed.

Each person on the team gets knowledge of the processes that the company follows. It also makes the role of each team member clear to comprehend.

Common mistakes that Project Managers make

Project managers are often guilty of making the mistake of micromanaging their employees. They seek control and empowerment away from their employees , resulting in the feeling of a lack of accountability. Some employees may feel that they do not have to follow the procedures to reach the end goal.

Each team needs to ensure it's self-organized. Members of the group should organize their activities based on their ultimate goals. They shouldn't be looking at a sequential path. In the event that a supervisor has to micro-manage his team, he's taking up his time and team's time.

What can you expect to gain from making use of online tools for project management?

If a project manager utilizes online tools, he'll be able to determine who is working or unemployed and evaluate the state of every project or process. These tools can reduce time as they can give the manager an idea of how work is reviewed, shared or discussed.

It is recommended to utilize online tools, such as Kanban as the project manager will reduce the amount of time spent in meetings and communications. There are many businesses that choose to use online project management because they have large teams spread across multiple places.

What makes Kanban so efficient?

The simple nature of the tool is what makes it efficient. Managers and teams can write all tasks on the backlog in the section for tasks on the board and select projects that must be a priority in the present moment. They don't need to be concerned about missing certain tasks due to the work load.

It also gives you a sense of accomplishment and satisfaction after cards are moved from the list of to-dos and onto the completed listing on the table. It is an easy method to see the tasks completed as well as the ones which are not yet completed.

What is it that makes Kanban adaptable?

Kanban instruments can be utilized in a variety of departments and different ways as it is an extremely simple tool. There aren't many processes that Kanban will not be useful. Many organizations use processes that are envisioned to make a busy day more manageable and organised.

The manager as well as the other team members can organize their tasks and determine if the work is under control by using this Kanban board. In the event that the leader is feeling stressed all he or she has to do is glance over the screen and settle down.

What is the best way to use Visual Thinking help in planning?

The saying "A photograph will speak a thousand words" is the reason it is essential to visualize the process. It is much more efficient to visualize the tasks that need to be accomplished in the day written out on the board rather than consider the tasks that have to be accomplished and then not remembering the majority of them during the course of.

It is thought that the majority of people are visual even when they do not intend to. Visual thinking is a great way to organize and create an order to the chaotic and unorganized thinking processes.

Is there a trend toward the use of visual project management generally?

This is because the majority of firms, regardless of size, value the benefits visual management and planning can bring to an business. This is apparent when looking at teams that are smaller. When teams that are smaller within an organization start to work efficiently, the company is likely to look for teams to utilize visually-based project management. As knowledge about Kanban is spread across the marketplace it is becoming more popular with those who want to implement the Kanban tool into their work.

If your team is trying to improve productivity, look at these points below:

Do not multitask since it can be an obstruction. There are instances when you'll need to be multitasking, as it is logical to do it.

Make sure to take breaks because you'll feel more relaxed and be able be more productive.

Make sure you finish those tasks that are the hardest to complete first. It is recommended to plan an entire day that is more manageable by the hour.

Switch off your phone or change it to silent when you're working.

* Do not schedule more than one meeting.

• Prioritize tasks on a regular basis and make sure that the sequence of tasks align with the requirements of the company.

Chapter 15: How Kanban Reduces Risk And Creates Improved Software

It's amazing how an idea that was developed by a grocery store and then modified to fit the car manufacturing industry can assist software development companies to create high-quality products that are less risky however, it does work. The most significant change is that instead taking physical material from bins it is the Kanban flexible project management system enhances the productivity of the organization within the software environment. The work tasks will be "pulled" to the pipeline of work when they are needed. Forecasts and schedules aren't the only thing that "pushes" the work into manufacturing.

The risk of the software system through:

* Reducing WIP (or work in process)

* Each stage of development is openly observed

The predictability of an organisation is enhanced through reporting and metrics

A minimal impact can be derived from the steps to change that are gradual, evolutionary and small.

The capacity to manage self is built by stimulating and increasing opportunities for teams

* The actual handling of tasks and understanding of processes for work is developed by the team.

Risks and challenges faced by the team are addressed in a rational and objective manner.

Software teams carry out these tasks using an Kanban procedure.

Visual work is the norm.

The board is a tool that is used to display the different steps of the process on an individual level. The use of this tool can help identify obstacles and issues before they escalate into a massive "fire." On a basic level the board is comprised of three columns or lists. They are "to do,"" "doing," and "done." When they are displayed in columns it is easy for the team to discern what remains to be done at each stage. This method simply displays the development of a project without the need to update an individual stakeholder by manual methods such as a phone telephone call or an email.

Visualizing the work helps the process of organization through:

* Breaking down each step beginning with "A" through "Z"

Each step has the column, or list for a smoother pipeline's execution and drawing lines

A simpler and quick-to-read monitoring of work is because of the color-coding used to identify different kinds of work when employed

* The status of the project is posted in central locations to inform all those

interested about the development of the project

Cards can be moved from column to quickly, and team members can be updated in real-time. This allows them to quickly address any issues or issues early.

WIP is not as extensive

Multitasking and wasting time is a bad idea in an agile work environment. Multitasking at once exposes the developer to making mistakes Each deliverable takes longer to complete and the process for delivery is longer. The restrictions set for the WIP means it is not possible to add any work until something has been removed, and the amount is set in accordance with the capabilities and capability that the group can handle. If each stage can take on more work, it's taken from a different location instead of being pushed to the to the top of expectations.

Projects are broken down into smaller parts and then each one is dealt with individually through the Kanban process. This ensures that workflow is steady and fast. The pipeline is clogged with limitations with a valid reason. If something blocks the system the team is charged with "cleaning" the blockage prior to adding new tasks to the pipeline.

Limits on WIP help with organizational throughputs by:

• Managing time by establishing a solid system and structure. Whatever the size of the team, or how complicated or straightforward the task it is a consistent procedure for each task that must be completed within a specific time frame.

A smooth workflow can eliminate the need for waste from your time as well as resources and costs. Additionally, it helps eliminate tasks that are not needed. This is the main purpose of WIP limitations.

Changes are incremental

The incremental changes that you make build on your current strategies in order to continue moving your project ahead. Recognizing the joints within the system that trigger back-up work is another advantage of using a Kanban system. WIP limits are set in the beginning to finish work in a short time and to inject new work into the system in a controlled fashion. As work progresses into each stage, the limitations set for WIP are refined and the phases shift as well as the enhancements to the workflow become more apparent. "Evolutionary" can be the word used to describe the Kanban method due to the way in which the method of work is done in smaller, bite-sized pieces.

Incremental changes increase throughput:

Continuous improvement is made every step of the process. Through to that final "done" phase, outputs are high-quality and outputs that are final are less susceptible to mistakes.

Flow Increase

In a conventional context, developers may not have the knowledge of what to do after the conclusion of the task they were given. The reason for this is that the previous work was impacted by an error that was found. If this is the case it is because the developers don't know what their next steps should be since the work is considered to be finished and they've already pulled up new work and changed the board. Implementing an effective Kanban system can prevent this from happening since the work job that is added to the board and then placed on an orderly manner is the most priority task on the backlog.

The process is designed to reduce the confusion and crisis that often happens when there is an inaccessible resource. "Done" work will only occur once the product has been finished and is in use. This method can reduce the time spent and the amount of work required to be re-worked.

Companies that adopt the Kanban method find that the effort and time spent in the process of adoption is well worth the effort. The difficulties of implementing an agile new system such as this can yield many advantages, including less risk and a better outcomes. Implementing the strategies outlined demonstrate how Kanban aids teams working on software development, and may be spread all over the world, but on the same grouping. This is possible thanks to virtual boards as well as integrated Kanban software. They provide easy-to-read WIP and strategies to offer system for tracking project status along with individual task assignments. The tools are designed to help make the use of a Kanban system more efficient on your employees.

Processes for development also benefit by the use of an Kanban system. It is possible to identify bottlenecks, and workflows are efficiently controlled by this process. If you're looking for a faster and more efficient workflow in your software development the Kanban procedure is an effective solution for your requirements. You can now offer top-quality software to your customers in the fastest time for delivery while reducing the risk of the process.

Chapter 16: Kanban Vs. Other Methodologies

Software development has employed various project management theories for years and this is a reason to ask what is the reason you prefer an Kanban method over other methods like SCRUM or resources similar to a Gantt charts? For starters, you must examine the differences between the principal methods.

VS Scrum

SCRUM is an extremely popular method of managing projects employed in software development today. Although it's a successful method of project management that is agile and flexible however, there is a significant distinction between the Kanban method and SCRUM.

For instance:

* The Kanban system doesn't include time boxes, as SCRUM calls for.

* The Kanban system can handle fewer tasks and are more extensive that SCRUM tasks.

*A Kanban system doesn't assess the processes often even if it does, like it would in the SCRUM environment.

* The Kanban system will only take into account the average time of completion for the project, instead of basing the project's time upon what is known as the "speed that the group" as in the SCRUM setting.

For those who are accustomed to working in the SCRUM environment, and believe that the project is comprised of team members' speed, greater size, and scrum meetings could find the thought to eliminate them completely absurd. These are the main method of controlling the development process within a SCRUM-based system! The issue with this idea is that it creates the impression of being in control. Managers constantly strive for control however the truth is that they'll not ever achieve it. Management's influence and supervision only function if the team will be willing to cooperate. If the team decides collectively not to press for the project's success no matter how the manager handles it; the project will fail.

Imagine a completely different scenario that is one in which people are having fun while working and are productive. Managers do not have to control the work environment. In this scenario, control would create chaos and increase the cost. In the SCRUM setting control measures can increase costs because they require constant discussions or meetings and also commitments to time during the transition of sprints. The majority of sprints need one day to finish and a day to begin the next. These extra days can be seen as "wasted" chance. If you view the situation as a percentage two weeks, that's 20 percent of the time to be spent making preparations and wrapping up. This is a lot of time! In certain SCRUM environments where up to 40 percent of the time could be devoted to supporting the methodology rather than in completing the task.

The Kanban method, however concentrates on the jobs. This is different from the SCRUM procedure. SCRUM professionals want to have the sprint to be successful. Kanban practitioners are looking for completed tasks. The tasks in the Kanban system are planned from beginning to end without an exact time frame for sprints. The finished work is presented and the project is released when it is completed. The tasks do not have an estimated deadline that is set in the hands of team members. This is because there is no reason to use the time estimation and estimates are generally incorrect. If a manager believes in and believes in their team, why is an estimate needed? Doesn't the team desire to do the best job they could in the quickest time? Instead, the manager focuses his or her attention on creating an inventory of tasks that are prioritized. The team's goal is getting as many jobs done as possible within the pool. It's as simple as that. The need for control measures is not necessary. Managers can add things to their pool and change the priority of tasks according to the need. This is the way an experienced Kanban practitioner manages an software project.

Conclusion

Kanban is a system of organization that was developed by Kanban system was invented by Taiichi Ohno in the context of Toyota automobile in the early 1940s. The system was later developed and implemented in the hands of David Anderson in 2004 in an IT firm. Following that, several companies from various sectors began using the system to enhance their operations.

With Kanban systems, organizations can pinpoint the different steps required in order to reach the ultimate objective. The board lists the processes involved and separates them into different levels they're in. This helps the team or members to spot potential bottlenecks and discover ways to address them.

In the course of the book, you'll discover the ways in which Kanban can be utilized by various departments to increase the quality of their products and deliver them in time. Kanban systems can be utilized to calculate estimates of delivery times for products using certain statistics tools.

In conclusion to conclude, it is clear that the Kanban system is a highly effective toolthat helps ensure transparency and give the company a better knowledge of how work is moving across departments.

Thank you for buying the book. I hope you've found the information you've been looking for.